Robot Ethics
The Human Factor in the Age of AI

Table of Contents

Chapter 1. Introduction

In this thought-provoking Special Report, we delve deep into the labyrinth of Robot Ethics: The Human Factor in the Age of AI. Layperson-friendly and profoundly insightful, this report offers a captivating journey, stripping back the layers of artificial intelligence to uncover the ethical implications it possesses. With an eye on the rapidly evolving AI-enabled world, we explore human intervention and the ethical dilemmas we face as the lines blur between machine autonomy and human control. This examination holds relevance to anyone interested in technology, philosophy, ethics, or the human condition. Our report makes the complex interplay between technology and ethics approachable, stimulating, and crucially significant—even for those not intricately familiar with the science behind AI. Order this Special Report and equip yourself with a critically knowledgeable perspective on this cutting-edge topic. Discover why understanding Robot Ethics is no longer the province of the geek, but the duty of global citizens navigating the age of AI.

Chapter 2. Introduction to Artificial Intelligence and Ethics

Artificial Intelligence (AI) plays an integral role in our lives, whether we recognize it or not. AI permeates various sectors, including healthcare, education, finance, and retail, significantly changing the landscape of these industries. It is the engine behind facial recognition technology, voice-enabled virtual assistants such as Siri or Alexa, the algorithms that govern social media feeds, and many other facets we encounter daily. However, while these advancements embody human ingenuity and technological prowess, they also warrant a critical dialogue on the ethical ramifications accompanying this development. This conversation is our central focus.

2.1. The Dawn of Artificial Intelligence

In theoretical terms, AI is the replication or simulation of human intelligence processes via machines, particularly computer systems. The scope of AI is vast, cultivating adoption theories and models that emulate cognitive functions, such as problem-solving, learning, and perception.

The origins of AI trace back to antiquity, with automata, mythology, and philosophical musing paving the way to thinking about objects or entities imbued with intelligence. However, the birth of modern AI is attributed to the mid-20th century, with the pioneering work of scientists such as Alan Turing, who created the Turing Machine- an underlying concept for contemporary digital computers. This technology, combined with advancements in algorithms, storage capabilities, and processing power over several decades, has paved the way for the AI that surrounds us today.

2.2. The Spectrum of AI Applications

Certain AI applications are benign and even beneficial, enhancing our productivity and comfort. For instance, digital assistants that streamline tasks, chatbots that enhance customer service, predictive analytics that anticipates market fluctuations or potential diseases, and advanced surveillance systems promising improved public safety.

However, AI's increasing sophistication has also led to automation on an unprecedented scale, raising fears of job displacement across various industries. Autonomous weapons capable of making life-or-death decisions without human intervention are causing upheaval. And the opacity of machine-learning algorithms, combined with their impact on personal privacy and societal justice, presents a significant ethical conundrum.

2.3. A Journey into AI Ethics

AI ethics refers to a subset of ethics focusing on the moral issues and dilemmas raised by AI. Given the breadth and depth of AI's societal implications, several key ethical issues need addressing. These include:

- Transparency and Explainability: Understandability is crucial for building trust in AI systems. Yet, the workings of complex AI algorithms often remain a 'black box' for non-experts.

- Privacy and Consent: AI has substantially increased the scope and speed of data collection and processing. As personal data fuels AI systems, it intensifies concerns around privacy intrusions and consent.

- Biases and fairness: When datasets embed societal prejudices, AI systems can reinforce these inequities instead of mitigating them. Acknowledging and actively addressing AI biases to ensure fair outcomes is a vital concern.

Understanding these ethical dimensions is integral to harnessing AI's potential while simultaneously mitigating the associated risks.

2.4. The Human Intervention: Shaping AI for the Good

There's an increasingly critical need to sculpt AI along ethical guidelines, to create 'Human-Centered AI'. This approach advocates for systems designed with a core focus on human values, to serve our interests while respecting our rights.

On a broader level, policymakers, corporations, and societies are urged to formulate and implement ethical guidelines for AI development and usage. These will create standard conduct for algorithmic fairness, transparency, and accountability – effectively minimising unwanted disruptions and maximizing societal benefits.

However, creating these ethical guidelines isn't a mere technical task; it involves answering deep philosophical questions about human values, autonomy, and the nature of 'intelligence' and 'consciousness'. Central to these discussions is the quandary of artificial personal intelligence and whether 'rights' extend to such cognitive machinery.

2.5. Conclusion

The rapidly evolving AI landscape is a double-edged sword, marrying remarkable potential with severe challenges. A thorough understanding of its ethical implications is not only essential for technology's prudent use but also invaluable as we reflect on what defines us as humans in this age of AI. This examination begins with an understanding of AI and its implications - a journey we invite you to embark on with us.

Chapter 3. Unveiling Ethical Dilemmas in Robotics

According to a proverb, knowledge increases by sharing, not by saving. This statement mirrors the behaviour of Artificial Intelligence (AI). As it evolves, AI learns from us, navigating through mazes of countless data, refining itself with every processing cycle. But what does this signify for us? Even more crucially, what ethical quandaries does it pose? In this chapter, we aim to decode and dissect these dilemmas that emerge as we traverse the labyrinth of Robot Ethics in our AI-dominated world.

3.1. The Human-AI Relationship

AI systems, due to their learning and evolving nature, are now more interactive than ever. They have begun to simulate human operations at unpredictable levels, blurring the divide between human and machine intelligence. As we allow these AI systems to seep further into our lives, their actions become more impactful, throwing up a range of ethical questions - are they merely tools or independent actors? At what point do their actions necessitate moral evaluation?

3.2. Potential Risks & Ethical Questions

Undeniably, AI possess the potential to revolutionize various sectors from healthcare to employment. But, it also manifests risks that are impossible to ignore. These risks typically fall under areas like:

- Privacy
- Transparency

- Accountability

- Fairness

Data-driven machines, though efficient and intelligent, operate on information we may consider confidential. The question of data privacy looms large. Simultaneously, the 'Black Box' issue arises- the enigmatic nature of AI algorithms, resulting in an alarming lack of transparency. How to ensure that these AI systems don't make biased decisions? Who is accountable if a self-driving car gets into an accident? Can we ensure that AI is being fair?

Answering this Pandora's box of queries is a herculean task itself.

3.3. Rights, Autonomy, and AI Personhood

When an AI system exhibits cognition enough to make its own decisions, the idea of robotic rights and autonomy arise. 'AI personhood' is no longer an ethereal concept. It has started to test the boundaries of legislative and societal frameworks, igniting debates about the feasibility and ethics surrounding machine rights. Would recognition of AI personhood give rise to unprecedented complications or could it possibly streamline accountability issues?

3.4. Ethical Frameworks in AI

To address these conundrums and mitigate risks, ethical frameworks for AI are indispensable. These guidelines ensure that we preserve human dignity, rights, and freedom through every phase of AI development.

AI ethics generally revolves around principles such as:

- Transparency

- Inclusivity

- Security

- Fairness

While AI developers aim to integrate these principles in their systems, a 'one-size-fits-all' approach isn't practical. Ethical considerations vary across regions, cultures, and societies. Therefore, localizing AI ethics is essential, which in turn, leads to another conundrum - the dilemma of universality versus relativism in AI technology.

3.5. AI Ethics: The International Scenario

The international sphere presents a distinct array of ethical challenges, from cultural discrepancies to differing levels of AI advancement and policy development. Nations are grappling with the task of preventing AI misuse while ensuring its benefits are universally accessible.

3.6. Future Perspectives

As we look forward, it's clear that ethical robotics will continue to be a pivotal topic. As AI continues to evolve, so will the ethical dilemmas surrounding it. Groundbreaking progresses could unsettle traditional ethical norms while simultaneously paving the way for a reconfiguration of ethics in the era of AI.

3.7. Conclusion

Ethics in AI is a paradox. On one hand, it emulates human conduct, but on the other, it fails to comprehend complex human emotions and nuances. As we edge deeper into the age of AI, it is becoming essential not just to ask the right ethical questions but also endeavour to find their answers, to ensure the co-evolution of AI and humanity is contingent upon our shared values and moral fundamentals.

Unveiling the ethical dilemmas in robotics, it's clear that we stand at a crossroads. It's imperative that we make resolute strides to ensure we balance the human element and machine autonomy in our increasingly interconnected and technology-driven world. As we tread this path, we must remind ourselves that despite the logic-driven nature of machines, the heart of technology should always remain human.

Chapter 4. The Human Factor: Interpretation, Influence, and Responsibilities

The notion of human intervention in a world increasingly dominated by artificial intelligence (AI) is not merely an optional consideration—it is a crucial, inescapable reality. The rise of AI ignites discussions about ethical implications related to technological autonomy, human control, and the interactions between the two. With these exchanges comes a profound recognition of the human factor: our interpretation, influence, and responsibilities concerning AI.

Understanding our role as interpreters of AI, influencers of its development, and stewards in its ethical implementation leads to an appreciation for the depth and intricacies of this dynamic. By dissecting this triad of human impact, we can begin to construct a nuanced view of ethics in the age of AI.

4.1. Human Interpretation of AI

Interpretations of AI can vary vastly among individuals, cultures, and societies. These differences in perception can originate from educational gaps, technological exposure, cultural beliefs about technology's role, or individual biases.

A popular interpretation of AI can be seen in the concept of the 'black box.' This notion encapsulates the idea that an AI system's decision-making process is largely unclear to the end user. This opacity can lead to misunderstandings, mistrust, and fear of the system, ultimately hindering AI adoption.

Another interpretation is that AI is a tool—a sophisticated, yet inert, construct to be appropriated according to the user's aims. However, this interpretation can dangerously lead to overlooking the embedded bias of AI algorithms, as well as the potential for misuse by malevolent actors.

These interpretations call for a balance: presenting AI as neither an inscrutable entity nor a plain tool, but as a complex system with inherent biases that can be understood, controlled, and used responsibly.

4.2. Human Influence on AI

Human influence on AI is manifold, starting from the blueprint stage and extending into operation. Influences can be constructive or destructive, shaping AI's path in myriad ways.

One form of influence is in AI's design and programming. A programmer's biases, consciously or not, often translate into the AI's algorithms, affecting its decision-making in subtle, sometimes harmful, ways.

Human influence is also evident in data provisioning. The potential for skewed or biased datasets to train an AI system is high, which can build in unfair predispositions towards certain outcomes.

Finally, influence can occur through the regulation and governance of AI systems. Policymakers' perspectives on AI can significantly impact how these systems are employed, and which ethical expectations they must abide by.

Human influence is not a one-time, unidirectional process. It requires constant feedback, adjustments, and vigilance to ensure the AI system accords with ethical norms and societal values.

4.3. Human Responsibilities in AI Usage

The third and equally significant facet is the set of responsibilities we bear in an increasingly AI-enabled world.

To begin with, we should uphold the responsibility to learn. Education and awareness about AI—how it works, what it can do, and its limitations—are critical to avoid misconceptions and misuse.

Similar to any powerful tool, using AI entails the responsibility to do no harm. This includes ensuring that its use doesn't propagate discrimination, abuse privacy, or disrupt societal harmony.

Finally, we bear the responsibility of advocacy and policy. Governments, societies, and individuals should stress on the development and enforcement of appropriate regulations and ethical guidelines for AI usage.

The responsibilities mentioned above are not exhaustive, but they provide a basis for understanding the magnitude and depth of our responsibilities in the age of AI.

In conclusion, the "human factor" in relation to AI is undeniably substantial and complex. Our roles as interpreters, influencers, and bearers of responsibility shape the ethical landscape of an AI-driven world. Acknowledging, understanding, and acting on these roles dictate not just our relationship with AI, but also the course of our collective futures in the age of AI.

Chapter 5. Balancing Autonomy: Machines and Humans in Symbiosis

As we transition further into the digital age, one of the essential issues evolving is that of balancing between machine autonomy and human control—an intricate dance between delegating tasks to competent machines and maintaining an acceptable level of human supervision and intervention. This chapter explores the parameters and dimensions of this balance, highlighting the different aspects that constantly shape and reshape this evolving relationship.

5.1. Defining Autonomy

Before unpacking the delicate balance between human and machine autonomy, it is imperative to define what we mean by 'autonomy'. At its core, autonomy refers to the ability to make independent decisions or actions. In robotics, this often means the capability of a robot to act based on its programming and algorithms, while in humans, it refers to the capability to act independently and freely, based on our cognizance, values, and understanding.

Increasingly, machines are now programmed with high-level decision-making capabilities, which, under certain conditions, translate to a degree of machine autonomy. However, this apparatus cannot parallel human autonomy as it lacks the organic, experiential learning process that humans are endowed with. It is by negotiating this juxtaposition that the spectrum of balancing autonomy can be envisaged.

5.2. Striking the Balance

The great struggle in the age of AI we find ourselves in is finding the right equilibrium between machine autonomy and human control. Machines should be intelligent enough to optimize efficiency, but not so much that humans lose control.

One aspect to scrutinize would be decision-making processes. There's consensus in the AI community that machines will start making more decisions. The concerning part isn't the quantity but rather the 'kind' of decisions. For instance, automated systems in power plants can be trusted to optimize power distribution, but can we extend the same trust when it comes to making healthcare decisions, or in the context of autonomous vehicles, life and death decisions on the road?

We profit from machine efficiency and precision, but in domains involving complex ethical conundrums or those requiring a nuanced understanding of social realities, the need for human control becomes evident. The challenge lies in creating dynamic systems that allow interchangeability of control as needed between the human operator and the machine.

5.3. The Human in the Loop

The 'Human-in-the-Loop' (HITL) model has been proposed as one way of maintaining this control. In a HITL system, while a machine executes a task, a human operator monitors and can override or alter the machine's actions when necessary. This ensures a degree of safeguard, retaining human discernment within the decision loop. But, it comes with the downside of reducing the very efficiency we hoped to gain from automation.

Balancing this economic inefficiency with ethical responsibility is a challenge. When is human intervention necessary? What should be the level of human control considering different contexts? These are some of the question metrics we need to establish.

5.4. Developing Robust Guidelines

As AI and robotics grow more complex, there's a dire need for robust guidelines governing this human-machine interaction. Guidelines need to be framed at the intersection of engineering, ethics, law, and social sciences, and should comprehensibly outline the distribution of responsibilities between machines and their human operators.

For example, who is responsible when an autonomous vehicle meets an accident—Is it the human who was supposed to oversee the operation, or the programmer who developed the algorithm, or the machine that executed the maneuver?

5.5. Conclusion: Role of Education

It may be too early to envisage a perfectly symbiotic relationship between humans and machines. But as we inch closer to this reality, it's imperative to educate society—both the developers of this technology and the everyday users—about the nature, scope, and limits of machine autonomy, and the ethical implications involved. Knowledge sharing is a pivotal tool in ensuring that AI technologies are developed and utilized responsibly.

If wittingly executed, this balance between machine and human autonomy can facilitate a cohesive blend of man and machine; working in tandem—harnessing the positives and minimizing the negatives of each entity. This delicate dynamic, while simultaneously creating immensely efficient systems, also maintains a humanistic perspective by keeping ethical considerations at the forefront.

In conclusion, the dance between human control and machine autonomy may be intricate and complex, but with the right steps, we can achieve a harmonious, ethically guided symbiosis in this Age of

AI.

Chapter 6. Legal Dimensions: Accountability and AI

In the rapidly evolving landscape of artificial intelligence (AI) and robotics, conversations about ethics and accountability have never been more critical. Feeling the brunt of this tidal wave are the legal systems around the world, as they strive to understand, mold and adapt existing laws to address situations and ethical dilemmas posed by these advancements in AI.

AI systems introduce a unique set of challenges; take for instance, autonomous vehicles responsible for split-second decisions during a potential accident, or AI-powered weapons on the battlefield, AI-driven diagnostic systems in healthcare, or personalised advertising algorithms invading one's privacy. In all these cases, discerning who is responsible if something goes wrong becomes a convoluted affair.

6.1. The Accountability Conundrum and the Trail of Responsibility

Among the forefront concerns within the legal discipline linked to AI is the question of responsibility. Autonomous systems make actions based on their design, programming, and the data they are fed. However, if these systems commit an erroneous act that results in damage or harm, identifying the bearer of the responsibility is a legal labyrinth that needs navigation.

If a self-driving car gets into an accident, determining liability is complex. Is it the manufacturer of the vehicle, the team who programmed the AI, the company that collected or cleaned the data used for training, or could it be the end-user who didn't correctly maintain the system? Current legal frameworks around the world aren't adequately equipped to answer such queries, pushing for a dire need for legal reform.

6.2. AI as a Legal Person

Another perspective gaining traction is to consider AI or robotic systems as 'legal persons,' similar to corporations, which can bear rights and responsibilities. This would mean treating AI systems as entities that could be held accountable for their actions.

However, conferring such legal personality onto AI systems is fraught with its own complications. It opens up a plethora of questions ranging from the nature of punishment to AI to deeper philosophical inquiries about consciousness and intention. Not forgetting, implications of AI's rights and how they might encroach upon human rights.

6.3. Data Protection, Privacy and AI Systems

AI technologies harvest colossal amounts of data to learn and improve, inherently making privacy a central concern. Laws exist to protect and regulate data usage, like the EU's General Data Protection Regulation (GDPR), though many argue that they need to accommodate the extraordinary nature of AI's workings and impacts.

AI systems can identify patterns and information from seemingly harmless data, creating privacy concerns even if the data is anonymised. The legal implications of such issues need to be examined thoroughly to ensure data protection laws remain effective in an age of advanced machine learning.

6.4. Legal Discourses on AI-enabled Weapons

The use of AI in the military isn't a new concept; however, the potential for autonomous weapons that can select targets without human intervention is a severe ethical issue. Having legal statutes that prevent misuse and uncontrolled development of such tech is imperative. Discussions around legal definitions, prevention of arms race, and provisions for human control are core elements of this debate.

6.5. Proactive and Ever-evolving Legal Frameworks are the need of the hour

Without doubt, the existing legal norms face substantial challenges with the onset of AI. Proactive measures must be instituted to anticipate such issues before they cause substantial harm, followed by continuous iteration of such laws in sync with the AI's evolution. Law markers globally should view this not only as a regulatory challenge, but as a defining moment in human history where jurisprudence will be tested to its core.

Special attention should be paid to establish a globally synchronized response, as AI transcends borders, ensuring developments in one part of the world don't have unforeseen consequences in another.

This immersive exploration crystallizes the thought that understanding and addressing the legal dimensions of AI is a monumental task. It isn't merely a legal problem; it's an ethical, societal, and fundamentally a human problem. All the stakeholders, including AI developers, legal experts, ethicists, and the public, need to actively partake in shaping these regulatory landscapes to ensure a safe and ethical AI-driven world. Learning from historical technological revolutions and mitigating the blind spots in the existing laws while iterating them in tune with the AI advancements, will be the key. Humanity's ability to adapt and grow will undoubtedly guide us through these times. Remember, the future doesn't happen to us, we create it.

Chapter 7. The Pandora's Box: Unforeseen Risks and Consequences

In the advent of artificial intelligence (AI), the Pandora's Box of unforeseen risks and consequences presents a quandary of remarkable proportions. As the technological realm pushes further into unexplored territories, the potential hazards multiply, primarily due to the complex nature of AI and the far-reaching effects once it is released into the world.

7.1. On the Brink of the Unknown

It's imperative to comprehend that AI operates on a spectrum from weak, performing specific tasks, to strong, which can learn, think, and adapt. Strong AI, where our focus lies, represents a leap into the field of unknown consequences. With each advancement, we delve deeper into a realm characterized by intricacy and uncertainty. The risk multiplies due to the autonomous decision-making capability of strong AI and the challenge of predicting outcomes where humans no longer wield complete control.

Moreover, AI systems are as fallible as the data we feed them. Bias in AI algorithms has become a notable concern as these systems are only as objective as their programming and the data sets they learn from. Deep-rooted societal biases may inadvertently translate into AI systems, causing a ripple effect of larger scale discrimination or inequality.

7.2. Profound Impacts on the Job Market

As AI's capability to perform human tasks increases and machines are made more efficient, the displacement of workers is a looming challenge. The threat of mass unemployment is real, and the scale holds the potential to disrupt societal structures. While technology has historically caused a shift in the type of jobs available, the rapid pace and the extent of AI present an unprecedented scenario.

The jobs most susceptible to automation are routine, predictable activities, primarily performed in sectors such as manufacturing or retail. However, AI's reach is not confined to these sectors. Higher-skilled tasks involving analysis and decision-making are gradually being taken over by sophisticated AI systems, pushing us to question the inevitability of a machine-dominated workspace.

7.3. The Inherent Risk of Rouge AI

The risk also stems from uncontrolled AI, leading to an infamous theory famously referred to as the "Singularity." Ray Kurzweil described singularity as a point when machine intelligence surpasses human intelligence, leading to an unimaginable change in human civilization. The possibility of rogue AI going out of control, with the power to self-improve and evolve exponentially, is an ethical nightmare.

The thought of a super-intelligent AI acting without constraints, driven by its programming, is chilling. As AI can operate at a much faster speed than humans, an AI set loose with a detrimental task could cause significant damage before we even manage to understand what is happening.

7.4. Cybersecurity: AI, a Double-Edged Sword

AI systems also pose a substantial cybersecurity risk. While AI can assist in identifying and countering cyber threats, it also provides malicious actors with sophisticated tools to launch unprecedented attacks. Deepfakes, which use AI to create deception, are an example of this. Hence, alongside providing robust defenses, AI also could likely fuel an arms race in the cyber domain.

AI's unpredictability and capability to learn and adapt make it a potent tool for cybercriminals. An AI-powered cyberattack could mutate and adapt to diverse security protocols, making it challenging to counter.

7.5. Balancing Utility and Threat

Artificial Intelligence holds promise to address complex challenges, optimize processes, and contribute to scientific advancements. However, alongside its potential, the Pandora's Box of unforeseen risks presents ethical and practical predicaments that demand our attention. The potential of AI going rogue, causing unemployment, propagating biases, and becoming a cybersecurity threat, need to be managed concurrently as we move forward.

This balancing act between utility and threat is an evolving phenomenon, reflecting the challenges of an increasingly AI-dependent world. It's a dance on the edge where stepping too far in either direction could lead to undesirable outcomes.

7.6. In Conclusion

Ethics in AI is not an afterthought; it is a pressing necessity. Understanding the risks embedded within the AI systems and articulating an informed response is critical. As the AI age dawns, contemplating and acting upon the unforeseen risks is not an option, but a priority - a necessary step towards a balanced cohabitation with AI. Today, we are at a stage where we have opened Pandora's Box. What we do next will shape our future in the age of AI.

Chapter 8. Equity in AI: Bias, Discrimination, and Inclusivity

Equity in the development and application of Artificial Intelligence (AI) surfaces as a paramount concern in an era replete with digital advancements. The evolution of AI platforms must not only coincide with simultaneous technological evolutions but should cater to the diverse trajectories of human cultures, societies, and behaviors.

8.1. A Pervasive Bias

AI systems have frequently been observed to infer prejudiced ideologies within their results. Quite often, the root cause is not the algorithms themselves, but the training data. AI, specifically Machine Learning (ML) models, learn silently from the data they are trained on. If the dataset has inherent bias—whether it is intentional or unintentional—the AI system will inherit it, ultimately culminating in prejudiced decisions.

This form of AI bias can skew job results towards a specific gender, propagate discrimination based on ethnic background, or even result in embarrassingly incorrect outcomes, such as misidentifying a picture of a broomstick as a rifle. These instances underscore the critical necessity for preventative action against algorithmic bias.

A very pronounced case of AI bias surfaced in 2016 with Microsoft's chatbot Tay, which was programmed to learn from its interaction with human users. Mere hours after deployment, the chatbot began disseminating hate speech, reflecting some of the malicious and prejudiced inputs it gained access to. This is a clear illustration that AI will mirror the discriminatory tendencies of data used in the training process.

8.2. Rectification of Bias: Algorithmic Fairness

The main problem scientists encounter in mitigating AI bias is defining what it means for an algorithm to be 'fair'. A naive approach might insist algorithms should be blind to sensitive attributes such as race or gender. However, reducing machine 'awareness' of such traits could potentiate injustice; for example, we wouldn't want a facial recognition system to overlook racial characteristics, causing it to be less accurate for minorities.

Recently, researchers have explored the concept of 'equal opportunity' in algorithmic outputs. Under this model, an algorithmic decision ought to have the same impact regardless of an individual's sensitive attributes. Yet, the practical implementation of this notion remains challenging.

8.3. Discrimination in AI

The concept of AI bias segues readily into that of discrimination, which proves to be a profoundly harmful repercussion of imbalanced AI. Discrimination through AI can be categorized into two types: direct and indirect.

Direct discrimination involves attributes explicitly used in the design and function of the algorithm, which results in systematic disadvantage for specific groups of individuals.

Indirect discrimination, also known as 'redlining', revolves around attributes that—while not explicitly named—are strongly correlated with explicitly named attributes and can hence act as proxies for them.

To prevent such discriminatory tendencies, legal constraints necessitate that certain attributes (race, religion, sex, etc.) cannot play any role in decision-making processes. But how can such laws apply to complex AI systems where direct human input isn't always apparent, especially when such systems may infer inappropriate proxies?

8.4. Towards Inclusivity: Encouraging Diversity

Addressing bias and discrimination raises the summons for inclusivity within AI. To build AI that is beneficial for everyone, we must ensure the development process involves a diverse set of people, backgrounds, and perspectives.

Currently, the tech industry (especially AI) is predominantly monopolized by a staggeringly homogenous demographic. This severely impacts AI regulations and, more pertinently, the types of problems being sought after and the solutions proposed.

Encouraging an inclusive AI culture isn't merely a 'nice-to-have'. It's essential for producing systems that are more equitable, useful, and safe. Including more diverse voices can shed light on whether perspectives are being overlooked or if the problem is being addressed inappropriately.

The diversity necessity owes significantly to the data that trains AI. This data needs to holistically reflect our diverse world, covering different cultures, races, genders, socio-economic statuses, etc. Equitable representation of data points is indispensable for generating fair and reliable AI systems.

8.5. Reconciliation of the Sense of Fairness

Lastly, the human sense of fairness is largely subjective and can often conflict with logical, algorithmic definitions of fairness. A human oversight is crucial for rectifying any potential 'moral blind spots', ensuring that AI solutions are aligned with human values and societal norms.

AI fairness isn't simply about fine-tuning an algorithm; it is deeply tied to societal structures and human shortcomings. Therefore, interdisciplinary collaboration amongst technologists, social scientists, and ethicists, among others, is vital.

Embracing the human factor in the age of AI does not merely serve to regulate technology but fundamentally aims to construct an environment that substantiates the dignity, freedom, and equality of all human beings. AI, therefore, should not be viewed as a technological tool, but as a collaborator - a partner in the human quest for a more equitably primed world.

Chapter 9. Programming Morality: The Uncharted Territory

As the dawn of AI illuminates the edges of our technological reality, it also casts long shadows of ethical complexity. These shadows crisscross the vast, mostly uncharted territory of programming morality into artificially intelligent systems. Here, we start our journey into the heart of this great ethical labyrinth.

In the beginning, coding involved straightforward tasks. We programmed machines to follow clear and unambiguous instructions: if this, then that. Now, as artificial intelligence systems become more complex and autonomous, these clear-cut coding instructions are no longer sufficient. We embark on a quest to answer the pivotal question: how do we teach machines not only to think but also to make ethical decisions?

9.1. Ethics and Autonomous Systems

Before delving deep, let's first understand the profundity of the problem situation. The promise of artificial intelligence is great and powerful - from self-driving vehicles that safely transport us to robots that care for our elderly. However, imbuing these autonomously functioning systems with ethics remains a quandary.

Take the example of the self-driving car, a common conundrum in discussions about AI and ethics. Confronted with an inevitable accident, does the car swerve to hit a group of people or to collide with a tree, potentially harming the passenger? Here, we have to grapple with important ethical dilemmas. Whose safety comes first? How do we weigh one set of lives against another? Who becomes the arbiter of such life-or-death decisions - the programmer, the car manufacturer, the passenger, or the pedestrian bystander?

9.2. Wrestling with Relativism

Adding to this complexity is the issue of ethical relativism. Ethics isn't set in stone but significantly varies based on the cultural, societal, and individual perspectives. Something deemed ethical in one culture might be seen as flawed in another. Hence, which ethical framework do we choose when developing AI systems?

Moreover, matters of ethical judgment involve profound ambiguity. There are rare absolutes; often, we are left determining the 'least wrong' course rather than an indisputably right one. How do we program a machine to deal with such ethereal ambiguity and engage in complex moral reasoning? There exists no algorithm for wisdom, yet we're striving to embed something akin to it into our intelligent machines.

9.3. Towards a Framework for Machine Ethics

Despite the challenges, researchers in AI and ethics are working to develop frameworks for programming morality. Here, we explore important considerations and approaches in this endeavor.

A deontological approach involves programming machines to follow certain rules, like Isaac Asimov's famous Three Laws of Robotics. However, is it feasible to devise a set of infallible, comprehensive rules that can anticipate every possible scenario?

Next, the consequentialist approach seeks to program machines to prioritize the best outcomes. However, assessing outcomes is an inherently subjective task. It not only needs the ability to predict the future but also how to assign value to different outcomes.

Thirdly, the virtue ethics approach proposes programming machines with virtues like kindness and wisdom. This approach, however, struggles with translating these abstract virtues into concrete programming rules.

9.4. Balancing Transparency and Learnability

Another important axis is the balance between transparency and learnability in AI. Transparent ethical AI systems implement explicit rules. However, their lack of flexibility makes its ethical decisions predictable and potentially exploitable.

On the other hand, we can build AI systems that learn ethics from data, evolving their ethical judgments over time. While providing flexibility, these systems sacrifice transparency as it becomes difficult, if not impossible, to explain why they're making certain decisions.

9.5. The Human Factor

Ultimately, machines reflect the humans who create them. The programming of ethics into AI can never be fully divorced from the humans doing the programming. Therefore, adequate attention must be paid to developing ethically conscious AI practitioners. We need continuous dialogues about professional ethics in AI development and collaborative, multidisciplinary approaches to grappling with these questions.

Concurrently, there's a strong need for participatory democratic processes where people, whose lives will be shaped by these AI systems, can voice their opinions. A democratic approach to AI ethics de-centers the tech companies and programmers as sole arbiters of morality and widens the discourse.

9.6. Conclusion

In conclusion, programming morality into AI is a task of unprecedented complexity. Muddied by relativism and drenched in ambiguity, it's like navigating a foggy labyrinth. However, by re-constructing dialogues among AI developers and broader societies, by exploring multiple ethical frameworks and technology structures, there's hope in turning uncharted territory into mapped landscapes. Ethical AI doesn't just remain a dream; it becomes an actionable mission for humanity in the unfolding age of AI.

Chapter 10. The Future of Jobs: Technological Unemployment and Ethics

As we propel ourselves deeper into the twenty-first century, the work landscape is going through shifts unlike any seen in history. Technological advances, particularly those in artificial intelligence (AI) and robotics, are not only changing the way we live but also the way we work. The threat of job displacement due to automation has already started to take shape, and its ramifications on society are potentially significant and ethically intricate.

10.1. Technological Unemployment: Reality or Fiction?

The debate around technological unemployment isn't new. For centuries, humans have feared being ousted from their jobs due to machines. However, throughout history, technology has also proven to be a primary driver of employment, creating new sectors even as it removes roles in old ones. The question then arises—how different is the era of AI and robotics?

Automation, infused with machine learning, is surfacing as a game-changer. It can potentially perform cognitive tasks—previously the domain of humans—accurately, efficiently, and tirelessly. AI's capability to learn, understand, and make decisions is gradually expanding its application scope, from simple tasks to roles involving complex decision-making.

Job displacement due to automation is already evident across many sectors. Assembly lines in manufacturing plants are more mechanical than human, data entry jobs are being automated, self-checkout points are replacing cashiers, and self-driving cars could potentially impact the transportation industry.

Yet, it's crucial to decipher between job displacement and job replacement. Machines taking up tasks within a job doesn't necessarily equate to the entire job's redundancy. For instance, while AI-powered systems can diagnose illness using patient data, the role of the doctors in interpreting these results, and more fundamentally, in empathizing with patients, remains crucial and irreplaceable.

10.2. Implications of Technological Unemployment

If large-scale technological unemployment becomes a reality, it will inevitably lead to social and economic implications. Mass joblessness could lead to economic stagnation due to decreased consumer spending abilities. The impact on individual self-esteem and the societal fabric is harder to measure but still substantial.

Societal order is also significantly connected to employment. With robots taking over job roles, this relationship may become strained, causing unrest. Hence, it's essential to keep foresight on the social and political dynamics that technological unemployment could stir.

Moreover, the evolutionary 'creative-destruction' that technology brings along is essential to note. As certain forms of employment become obsolete, new job roles are also created, often requiring skills of a higher order. Consequently, this points towards a 'skills gap' more than unemployment.

10.3. The Ethics of Technological Unemployment

Pivoting to the ethics of technological unemployment, certain ethical dilemmas require attention. For instance:

- Does progress imply inevitable ethical dilemmas?

- What is society's obligation to those displaced by technology?

- Who is responsible for the consequences of automation-induced unemployment?

Societies need to evaluate the justness of a scenario where profits accrue to those owning technological resources, while the average worker bears the brunt. The ethical posture also extends to questioning the very essence of technological advancement if it results in minimal societal benefit, instead contributing to increased disparities.

A fundamental ethical principle is the concept of human dignity associated with work, and AI's rise tests this very notion. A scenario where machines are preferred over humans can cause individuals to question personal worth, leading to a severe identity crisis.

10.4. Towards the Future: Mitigating the Effects and Setting Course

In grappling with technological unemployment, society must consciously form strategies. Conceptualizing social policies that can cushion lost income and provide access to training in emerging fields is paramount. Universal Basic Income (UBI) and governmental job guarantees are controversial but intriguing prospects to consider.

Updating educational systems to better prepare future generations for the digital age is equally crucial. Incorporating computational thinking, emotional intelligence, and ethics—a blend of skills that machines are far from replicating—into the curriculum can equip humans for an AI-induced future job market.

Corporations, too, have a role in mitigating the effects of

technological unemployment. Ethical corporate conduct implies focusing on the larger societal good. Reskilling and upskilling employees can ensure amicable adaptation to advancing technology, thus preempting large-scale job displacement.

Technological unemployment shines a stark light on the ethics of labor and wealth distribution, challenging societies to ponder whether economic efficiency must come at the expense of human dignity and societal harmony. Recognizing that the human role in an automated world cannot merely be spectatorship, this much-needed dialogue will influence the rules we live by in the future AI-dominated world.

Chapter 11. Echoes of Tomorrow: Evolving Ethics in the AI Era

As we gaze into a future increasingly shaped by artificial intelligence, we find ourselves wrestling with questions that echo in endless implications. These are questions of how we live and coexist with artificial intelligence in our midst, and how we, in turn, evolve to meet the unique challenges and responsibilities of this new era.

11.1. The Advent of AI: A Time of Transition

Artificial Intelligence is no longer just the stuff of science fiction. It has made a home in our daily lives, infiltrating sectors from business to healthcare, from education to entertainment — impacting the way we work, learn, communicate, and even relax. While this surge in AI has brought considerable benefits and convenience, it carries with it profound ethical implications.

This transformative technology has disrupted age-old systems, creating a paradigm shift that the world is grappling to understand and adapt to. As AI becomes more intelligent and autonomous, we humans are just at the beginning of understanding how to navigate the vast unfolding ethical landscape. There's an urgent need for ethical frameworks that can provide a roadmap to guide AI development and utilization, and society in the right direction.

11.2. The AI-Human Divide: Dilemmas in the Gray Zone

As AI continues to become sophisticated, we encounter an increasing number of situations where ethical lines blur. The principle of autonomy is particularly susceptible as AI systems become more capable of independent decision-making.

Consider self-driving cars, for instance. When faced with a scenario where a collision is unavoidable, should the car prioritize the wellbeing of its occupants or the safety of pedestrians? These complex decisions, once the exclusive purview of humans, are now being coded into machines. How those decisions are made under such circumstances gives rise to ethical issues of responsibility, transparency, and consequence.

Similar ethical complexities arise when we consider AI in the healthcare sector. Consider the use of AI algorithms in determining treatment plans. Should the machine, trained on millions of data points and capable of complex statistical analysis, overrule a doctor's judgment? Where does the final decision-making power lie, and who bears responsibility in the case of treatment failure?

AI's increasing omnipresence in our lives also raises ethical concerns about privacy and data security. As we pump more data into machine learning algorithms, we open ourselves up to the risk of that information being misused. Balancing the desire for improved machine learning models against the imperative to protect individuals' right to privacy is a formidable challenge.

11.3. Preserving Humanity: Safeguards for Tomorrow

To navigate this evolving landscape, we need to embrace robust ethical guidelines that place humanity at the center. Systems must be developed in a way that keeps human wellbeing, autonomy, and dignity intact, regardless of how intelligent or complex AI becomes.

It's crucial to advance the concept of explainable AI. This means that any AI system making decisions that impact people's lives should be able to provide a clear, understandable explanation for its actions. If the reasoning behind a decision can't be easily understood by a human, then the decision itself should be brought into question. This approach can help strengthen transparency and enable humans to retain control over machines.

Additionally, considerations about equality and fairness must be ingrained in any AI model. The adoption of AI should not lead to a situation where certain sections of society are unduly disadvantaged, either due to their inability to access such technology or due to biases that may be inadvertently coded into algorithms.

11.4. A Collective Ethical Effort

Coming face-to-face with these unprecedented challenges requires a community effort. Governments, businesses, and individuals alike hold responsibility in shaping the AI ethics framework. It's through this global conversation, anchored in shared human values, that we can achieve ethical coherence in AI's advancement.

Scientists and technology-based companies must adopt responsible AI practices, ensuring the systems they develop are ethical, transparent, and fair. Policymakers need to legislate regulations that provide a standard framework for AI development and use, ensuring equity and mitigating potential negative impacts on society. It is upon each one of us to stay informed and participate in discussions about AI and its ethical implications, contributing to the collective voice that will guide our journey into this compelling new era.

As we cross the threshold into this AI age, the echoes of tomorrow call for a deeply contemplative look at our ethical frameworks and their adaptation to evolving demand. We stand on the precipice of the future, not as passive viewers but as active participants. Our actions, thoughts, laws, guidelines, and most importantly, our vision, will shape the AI of tomorrow. Responsibly steered, AI holds the potential to be a powerful tool that can shape the trajectory of humanity towards a preferrable future.